WINDMILLS OF BIRMINGHAM
AND THE BLACK COUNTRY

WINDMILLS OF BIRMINGHAM
AND THE BLACK COUNTRY

by

Joseph McKenna

K.A.F. Brewin Books.

First published May 1986 by
K.A.F. Brewin Books, Studley, Warwickshire.

ISBN 0 947731 10 5

Typeset in Baskerville and
printed in England by
Supaprint (Redditch) Limited
Redditch, Worcestershire.

CONTENTS

ILLUSTRATIONS

INTRODUCTION

It is said that the windmill was introduced into this country by the Crusaders at the end of the 12th Century. At the opening of the 14th Century we find references to established windmills at both Walsall in 1304, and at Sutton Coldfield in 1309. These mills would have been the earliest and most primitive form of windmill, the Post Mill. The name is derived from the central post upon which the whole box-like structure revolves. Originally this post would have been sunk into the ground, but later was suspended above the ground by trestles. They in turn rested on stone or brick-built piers. Thus the wooden mill was kept clear of the ground to protect it from rotting. To turn the mill into the wind the miller was obliged to turn the whole structure by means of a long beam, which projected from the back of the mill. This beam was known as a tail-pole. Sometimes to facilitate easier movement a cart wheel was added to the end of the pole. Later, store houses, known, because of their shape, as roundhouses, were added beneath the box-like structure of the mill. Local examples of Postmills with roundhouses, include Wake Green Mill, Sandyfields Mill, and Essington Mill.

When the wind changed the miller had to stop work and turn the whole mill back into the wind. It was both tiresome and a tiring necessity. During the middle of the 16th Century the Tower Mill was invented in Flanders. As its name suggests it was a brick or stone tower surmounted by a cap and sails. Its advantage was that only the cap and sails revolved. Originally tower mills were turned by the traditional tail-pole, as at Chapman's Windmill at Holloway Head, but later by an endless chain gear mechanism. Finally came the fantail gear mechanism. The advantage of the fan-tail was that it was self-righting. The fantail was like a small windmill, set at right angles to the sails. As soon as the wind changed it caused the blades to revolve, driving the cap through a series of gears, to bring the sails back into the wind.

Though of more advanced design than the Post Mill, the Tower Mill was expensive to build. During the 17th Century a new form of Tower Mill, constructed of wood, came into being. It was known as the Smock Mill, from its fancied resemblance to a farm labourer's smock. Usually octagonal in shape, like the Tower Mill, the main body was static, with only the cap moving. The only confirmed sample of a smock mill in the locality is Cooper's Mill in Deritend.

In 1772 the scottish millwright Andrew Meikle invented the spring sail. It was composed of a series of shutters arranged like a Venetian blind. The blades were opened or closed depending upon the strength of the wind. Meikle's invention superseded the old common sails, which with slight variations, had been in use in this country since the introduction of the windmill. Common sails were merely wooden frames on which canvas was spread. Their great disadvantage was that the mill had to be stopped each time there was a change in the wind's strength, for the sails to be adjusted. In wet or very windy conditions this could prove a difficult task. William Cubbit improved upon Meikle's invention in 1807 with his patent sails. He retained

Meikle's sail, but drilled a hole through the windshaft. A rod was passed through the hole, and connected to all the sail bars on the four sails. At the junction of the four striking rods was a gear arrangement linked to an endless chain which hung down to the ground. Thus the miller could control his sails in one single operation without stopping the mill.

Finally the cap which protects the tower mill and its machinery. There are six distinct types of cap. The gable is triangular in shape, like the roof of a house. The postmill, which looks like a miniature mill surmounting the tower. The boat resembled an upturned dinghy. The cone, which was either circular, hexagonal or octagonal, was not as common as the other types, but there was one atop the Ruiton Mill at Upper Gornal. The domed cap was the fifth type, and finally the ogee cap, the perfection of windmill caps. Originally these windmill caps were regional in design, but travelling millwrights spread their designs nationwide.

BIRMINGHAM

It is difficult to imagine windmills existing in a modern city like Birmingham, and yet just over 100 years ago they were still a common sight. An image of stability in an already changing world. So taken for granted were they that both travellers and natives alike, never even bothered to write of them. And now they are gone, every single one.

Several of Birmingham's windmills were Medieval in origin. The earliest extant reference is to Maney Hill Mill in the north Birmingham parish of Sutton Coldfield. It was in existence as early as 1309. The last windmill to be built in Birmingham appears to be Holdford Mill at Perry Barr, in the parish of Handsworth. In this same parish, at a place called Soho, a new form of power was perfected. A power that was not variable like wind; a power that did not cease in times of drought. That power was steam. With its coming, the days of the windmill were numbered.

Within the old parish of Birmingham, including Deritend, there were four windmills, Cooper's, Chapman's, Baskerville's and Birmingham Heath mill. Cooper's mill was situated in Heath Mill Lane, Deritend. It is shown on Samuel and Nathaniel Buck's South-West Prospect of Birmingham, published in 1731. The Cooper family also worked the nearby water mill, shown on William Westley's East Prospect of Birmingham. In a view of 1779, Cooper's windmill is shown as a smock mill[1], a point confirmed in Thomas Dixon's view of 1826, where the mill is shown with common sails.[2] Development of the area about this time ensured the mill's demise. Its last miller was Thomas Whitmore, who held Cooper's mill from 1825 to 1836.

On the hill overlooking Holloway Head was Chapman's windmill. It succeeded at least one earlier mill on this site. Gordon Roe in his biography of the Birmingham artist, David Cox, writes, "Joseph Cox's first wife, David's mother, was born Frances Walford, the daughter of a miller and farmer who gave his name to the mill which he had built himself on a high gravelly hill near Holloway Head."

The usual date given for the construction of Chapman's brick-built tower mill is 1745. It was probably built by a member of the Clark family who were millwrights in Deritend until late into the 18th Century. Chapman was still in possession of the mill in 1774, when he made his will,[3] but two years later his executors sold it to John Griffiths. Griffiths put his brother Thomas in as miller while he ran the business as a baker and miller at 114, Hill Street.[4] In April 1778 the mill was struck by lightning. The main beam, one of the sails and a cog wheel were "shivered in pieces, and it was otherwise materially injured, the miller was struck down by the violence of the shock, and remained senseless for some time."[5]

William Martin took over the mill from the Griffith family, four hundred pounds having been expended on repairs. Originally the mill had a tail-pole, as is seen in an early engraving, but now this was replaced with a fan-tail gear mechanism. During

Cooper's Windmill

1820's and early 1830's Thomas Lucy of the Stratford-upon-Avon milling family worked Chapman's mill with his sons. After him came the millers Sandbridge, Kendrick and Bridges. With Bridges' death the mill fell into decay. The land surrounding it was laid out as a bowling green and pleasure garden. The mill was used as a camera obscura, and from time to time for firework displays under the direction of "Signor Castileo."[6] In 1873, with the aid of gunpowder, Chapman's windmill was demolished.

John Baskerville, the Birmingham printer of international fame, erected a windmill on his estate at Easy Hill sometime after 1745. It is shown as a small tower mill, devoid of its sails, in a view of 1789, by the artist David Parkes. The mill is described in Baskerville's will as a "conical building in my own premises heretofore used as a mill," wherein he left directions that he should be buried. The windmill was used by Baskerville in the production of high quality paper for printing.[7]

Birmingham Heath Mill was built on the common land to the north of the town. There by 1814, the windmill is shown in a painting by Samuel Lines as a squat, white-painted structure, in his view from St. Philip's cupola. It lay some distance back from the Dudley Road, approached by a fordrough, opposite to where Heath Street North now is. Joseph Flecknoe was recorded as miller in 1830. A smock-mill, it ended its working life in 1849 when the land was secured by the Guardians of the Poor for the erection of a new Workhouse.[8]

Northwards now, across the parish boundary into Handsworth. Soho Mill was situated to the north of Hockley Brook. Samuel Botham in his 1794 Survey of Handsworth records field number FK932 as "Little Windmill Hill." It is not shown on Joseph Browne's Map of Staffordshire for 1682, and was probably Medieval in origin.

East now to the Birchfield Road. On the edge of the Aston Hall Estate, was Hutton's mill, built in 1759,[9] for the Birmingham historian William Hutton. He used it in an unsuccessful attempt to produce paper cheaply. Cheated by his workman, or so he believed, he resolved on 30th June 1761 that he was "determined to make no more paper at the mill, but dispose of it, if I could, if not, convert it to another use."[10] Hutton eventually sold it to Rebecca Honeyborn in 1763. A bottle-kiln shaped tower mill, it was eventually converted into dwelling houses and became known as Birchfield Round House. Hutton's windmill was demolished at the end of the 19th Century.[11] Further along the Birchfield Road, just off the present Livingstone Road, was Bristnalls End mill. The mill had disappeared by the time Samuel Botham undertook his survey, but he recorded field number BJ320 as "Windmill Piece". This field and its neighbour, Berry's Croft, were offered for sale in 1813.[12]

Holdford mill was built sometime after 1794.[13] It was situated at the farm of the same name, near the "zig-zag bridge", just to the east of Aldridge Road. It is shown on the 1 inch to 1 mile Ordnance Survey map of 1834, and in the 1843 Handsworth Tithe, where the Rev. Dr. Henry Weedall [14] is shown as its owner. Possibly a post or smock mill, there is no trace of it on the first 25 inch to 1 mile map of 1889. The last of Handsworth's windmills, Hamstead mill, was situated south of the River Tame, between the present Hamstead Wood and the fish pond. "Windmill Hill is recorded on Botham's Survey of 1794. Long gone by that date, and indeed not even recorded in 1682,[15] the mill was probably Medieval in origin, and supplanted by the nearby watermill.

Chapman's Windmill

To the east of Handsworth lay the parish of Aston. Within this parish were three windmills in addition to Cooper's mill at Deritend. Bleak Hills mill was situated just east of Witton Upper Pools. It was built sometime after 1760, as it is not shown on the Erdington Estate map of that year. Certainly it was in operation by 1814, and in 1830 being worked by a miller named Oldacre. Elizabeth Oldacre, possibly his widow, was in possession in 1834,[16] though the mill was owned by the Handworth landowner Wyrley Birch. The mill had ceased operation by the end of the 19th Century.

Nearby was Slade Mill, situated midway between Slade Road and Witton Lower Pools. It probably dated from the Medieval period.[17] Slade Mill is not shown on any of the early 18th Century maps of the district, but is perpetuated on Fowler's map of Aston Parish, in "Windmill Hill."[18] The third of Aston's windmills, Saltley, is shown on Henry Beighton's map of Warwickshire for 1725. Samuel Buck's South-West Prospect of 1731 depicts it as a postmill. Saltley Mill is recorded on Thomas Jeffery's map of 1755, but not shown on Tomlinson's map of 1760. The site is marked as "Mill Field", and on Fowler's 1834 map as "Piece Mill Field". It is difficult to position the mill's site exactly, due to the heavy development of the area, but it would appear, basing the site on Fowler's map, to be at the junction of Saltley and Ralph Roads.

Sutton Coldfield, Birmingham's newest suburb, borders onto the old parish of Aston. The oldest recorded of Birmingham's windmills was situated here at Maney Hill. The mill is referred to in a document of 1309'[19] wherein it is recited that,
". . . none of these customary tenants used to grind his corn but at the lord's water-mill, so long as the mill was in repair for grinding, unless they first paid the whole of the corn to the lord's miller, upon pain of forfeiture of the whole corn; excepting the tenants of Maney, Windley and Wigula, who ground at the lord's windmill at Maney."

Probably the first of several windmills on this site, one of the four pier stones on which one of the later mills rested was discovered in the early 1850s:
". . . a large stone was turned out of a hedge-row on the hill; it measured about five feet in length and two feet in width and thickness and was of a fine grained, hard, dark, substance, apparently limestone or trap; but it was unfortunately broken up for the roads before its nature could be ascertained."[20]

There were two other windmills at Sutton, Langley Mill and High Heath Mill. Langley windmill was probably run in conjunction with the nearby watermill. Possibly of 17th Century origin or earlier, this windmill is not shown on any of the early 18th Century maps, and was most certainly gone by 1722.[21] The site is clearly marked as "Windmill Field" on the Ordnance Survey 2 inch to 1 mile drawing of 1811, and confirmed as such on the 1 inch to 1 mile map of 1834.

High Heath Mill was situated just under two miles east-north-east of Sutton Coldfield town centre, between the Tamworth and Withy Hill Roads. Built by 1725, this postmill is shown on both Beighton's map of 1750, and Jeffery's map of 1755. It was demolished though by 1788 when James Sherriff was surveying for his map of 25 miles round the Town of Birmingham.

Hutton's Windmill

Within the parish of Yardley there were three windmills.   Yardley mill, alternatively known as Greet or Red Hill Mill, was situated just south of the Coventry Road near the former Adelphi Cinema.   The earliest evidence of a windmill here appears in the Inventory of "Jone Wall", drawn up in 1579, which refers to a "Windmyll Filde."[22]   The mill was being worked by Henry Barrs in 1668-9,[23] and by Richard Barrs in 1725,[24] where Beighton's map shows it as a tall postmill.   The mill was still in use in 1755,[25] but by 1800, was known as "Old Mill."

Wake Green Windmill

Further south, where Hall Green forms the parish boundary with nearby Moseley, was Wake Green Mill. A postmill, situated on a small knoll in what is now the playing fields lying just off Yardley Wood Road before its junction with Swanshurst Lane. The mill is referred to in a deed of 1664[26] when it was in the possession of Richard Grevis. The deed relates to "all that windmill with the appurtenances situate and being near unto the said Watermills (at Greethurst and Swanshurst)". In 1766 John Allen leased the mill, then in the possession of John Taylor, for 14 years at an annual rent of £18. He was still in possession in 1773 as is shown by an indenture of that year which refers to "All that windmill with its Appurts situate near a place called Moseley Wake Green . . . standing upon part of the common or beaste land belonging to and being parcel of the Manor of Yardley."[27] David Cox, the Birmingham artist, immortalised the mill in a sketch of 1819. Sometime between that date and 1834 the mill was demolished.

The third windmill was situated in the south of the parish, at Yardley Wood. Not to be confused with the tower mill just across the border in Solihull Lodge. This postmill was built in what is now a playing field near Christ Church. It is shown pictorially on Taylor's map of Worcestershire for 1800, its site is marked as Bach Mill.

To the east of Yardley village lay the parish of Sheldon. Of its three mills, two were probably Medieval. The oldest appears to be Lyndon Green mill. It is not shown on any of the early 18th Century maps, but its site, now bounded by Elmay, Benedon and Larne Roads, is recorded in the Sheldon Tithe of 1840 as "Windmill Close." Wells Green Mill was situated north-east of the Green. It is depicted as a postmill in 1675.[28] Demolished very early in the 18th Century, its site is marked as "Old Mill Hill" on both Beighton's map of 1725 and Jeffery's of 1755. Sheldon windmill, just off the present Rectory Park Road, was apparently built to replace it. In existence by 1788,[29] on or very close to the site of the old Wells Green windmill, it is shown as a postmill. In the Sheldon Tithe of 1840, the mill is owned by Richard Greenway, and being worked by John Greenway. Interestingly two fields in close proximity are each known as "Windmill Close," and may mark the site of the earlier Wells Green windmill.[30]

In Edgbaston, noted more for its watermills than its windmills, there was one, run in conjunction with a much older watermill. In 1810 John Heeley was paying rates for "the two Speedwell mills."[31] The windmill is shown on a map of 1814,[32] and on Greenwood's Map of Warwickshire for 1822, where it appears to be known as Edgbaston Mill. Members of the Stratford-upon-Avon family of Lucy were millers here in the late 1820s and early 1830s. The mill is not shown on the Ordnance Survey Map of 1834, and was presumably no longer in existence.

Within the parish of King's Norton were two windmills. Situated to the north of Weather Oak Hill and at Headley Heath, they both now lie outside the city boundary. Northfield parish also had two windmills. Within the sub-manor of Haye and Middleton was Middleton Hall mill. It's site is recorded as "Windmill Hill" on the Northfield Tithe of 1839.[33] Demolished sometime before 1722, it was probably Medieval in origin, as was the parish's other mill,[34] Shenley Fields Mill. Documentary evidence of 1692 refers to it as being in Upper Shenley Fields in Shenley Lane.[35] There is no trace of the windmill on the Tithe map.

The last known of Birmingham's windmills was in the parish of Quinton.

Ridgacre mill was situated in the circle of roads that now make up Hagley, Clydesdale and Ridgacre Roads, and Stoney Lane. The site is recorded in the Ridgacre Tithe of 1844 as "Windmill Piece," and on the 25 inches to 1 mile Ordnance Survey Map of 1883 there is, a little way to the east, Windmill Farm. Most probably a postmill, it is not shown on any of the 18th Century maps and appears to have been demolished before 1722.

1. East View of Birmingham in Warwickshire. Engraved for the Modern Universal British Traveller, 1779.
2. View of Birmingham, Engraved by T. Dixon and published by Fisher, Son & Co., Caxton, London, 1826.
3. Pearson's Newspaper Cuttings, vol. 1. p.179 (B.R.L. 310039)
4. Pearson & Rollason's Directory of Birmingham 1777.
5. Gentleman's Magazine, April 1778
6. Pearson, p.144, 179, 208, 311.
7. W. Bennett, John Baskerville vol. 2, p.125.
8. Pearson, vol. 1. p.71 & 80.
9. B.R.L. 324357.
10. William Hutton, "Recollections" B.R.L. 495245.
11. See also: Pelham, R.A., "Hutton's Paper mill and its geographical significance. T.B.A.S. LXVI (1950) p 150-5, & McKenna, J., "Mr. Hutton's Windmill," T.B.A.S. 88 (1976-7) p.136-8.
12. Aris's Birmingham Gazette, Monday 24th May 1813, p.2. The accompanying plan is now housed in the Local Studies Dept. of the Central Library, Birmingham (B.R.L. 304039).
13. Not shown on Samuel Botham's Survey of the Township of Handsworth, 1794.
14. Handsworth Tithe 1843, Field No. 1060.
15. Joseph Browne's Map of Staffordshire, 1682.
16. Fowler's Map of Aston Parish, 1834.
17. W. Seaby, Warwickshire Windmills, p.4.
18. Plan Nos. c264 & 271
19. William Dugdale, Antiquities of Warwickshire, 1730 p.911-2.
20. Agnes Bracken, History of the Forest & Chase of Sutton Coldfield.
21. Beighton, 1722-5.
22. Worcs. Record Office 008-7/120/1579.
23. Warwick Record Office CR299/285.
24. Yardley Court Roll, 1724 BRL 434743.
25. Thomas Jeffery's Map of Warwickshire.
26. B.R.L. Meath Baker 438.
27. B.R.L. 324328
28. John Ogilby, Britannia, 1675 plate 74. Thomas Cookes was miller.
29. James Sherriff's Plan of upwards of 25 miles round the Town of Birmingham, 1788.
30. Field Nos. 560 & 554, Sheldon Tithe, 1840.
31. Edgbaston Rate Book, 1810 B.R.L. 345255.
32. Ordnance Survey 2 inch to 1 mile drawing, 1814.
33. Plan No. 256.
34. R.P. Hastings, Discovering Northfield, p.16.

In 1686 Dr. Robert Plot's History of Staffordshire was published. To accompany it, Joseph Browne prepared a map. For the windmill enthusiast it is a most wonderful map. There is an abundance of mills that leap out at you, and no more so than in the region that was to become the Black Country.

Smethwick Windmill was situated in a yard near the corner of Ballot Street and Windmill Lane, opposite the Windmill Inn. The land whereon it stood was purchased in February 1803 by William Croxall, a miller, from John Reynolds for £37. A 50 feet high brick tower mill with 3 pairs of stones, the mill ceased working in 1873. Edward Cheshire bought the site with its mill for his brewery, and the mill featured prominently in its advertising. The Brewery itself adopted the name "Windmill Brewery". Sadly though the mill was only used as a storehouse for barley. Late in its life Smethwick Windmill was used as a bicycle park and finally a Civil Defence lookout. In August 1949 the mill was examined and found to be structurally unsafe. It was demolished, but in the process the iron boss at the end of the main windshaft which held the sails, was shown to bear the date 1707.[1] It seems probable that Croxall incorporated the remains of some earlier mill into his windmill at Smethwick. The sails, as well as a portion of the mechanism which had been removed some years previous, are now exhibited in the Science Museum in South Kensington.[2]

North-west of Smethwick lies West Bromwich.   It possessed four, possibly five, windmills.   The manorial windmill, at Hall Green, was probably worked in conjunction with the nearby water mills.   It was situated just south of the present Hall Green Road.   The mill was in existence by 1616.[3]   In 1673 all the manorial mills, including the windmill, were leased to Jeremiah Parker.[4]   Joseph Browne shows the mill on his map of 1682, as does William Yates on his 1775 county map.   It was there still in 1816 as the 2 inch to 1 mile Ordnance Survey drawing shows, but was gone by 1834.

Sandyfields Windmill

There was a second windmill at West Bromwich. Wood's Plan of West Bromwich for 1837, shows it at the junction of the present Mill Street and Tantany Lane. The Ordnance Survey Map of 1834 does not appear to show it, possibly due to the smallness of scale and build-up of the area. White's Directory of Staffordshire for that same year does however list Charles Higgins as miller. Mary Willett confirms that the mill was still in existence in 1842,[5] but it appears to have been demolished soon after. Lyndon Mill, the third of West Bromwich's four known windmills, was situated in the triangle of roads that now make up Church Vale, Tenscore Street and Little Lane. The mill's site is shown on a plan of the Hallam Estates, drawn up by R. Court in 1808.[6] Possibly due to the heavy development of the area, the mill is not shown on either the 2 inch to 1 mile drawing of 1816, or the completed Ordnance Survey map of 1834. Various directories of the 1830s appear to refer to this windmill as the corn-mill at Churchfield. Lyndon windmill had been demolished by 1889, as it is not shown on the 6 inches to 1 mile Ordnance Survey map of that year. The last of West Bromwich's windmills was Rider's (or Ryder's) mill. Built by the family of that name on their estate to the east of the town, sometime after 1682.[7] Some twelve years later a mortgage of 20th May 1694 refers to "Answorth Leasow, or Windmill Field, with the windmill erected thereon."[8] The mill is shown on a plan of 1767, south of the River Tame,[9] but had gone by 1775.[10]

Walsall too had its share of windmills, including the oldest known in the locality. In 1304-5, "Thomas le Rouse, Knt., Lord of a Purparty of Walshale," granted to "Sir Roger Morteyn, Knt., and his heirs, a reasonable road to his windmill of Walshale..."[11] The following year Sir Roger granted the mill to Henry de Prestwode and his son John.[12] In 1318 John sold his interest in the mill to Ralph, Lord Basset.[13] Caldmore mill apparently passed into the hands of the Lord of Walsall, and was held by him and his successors from 1385 to 1393. In that year it was destroyed by a gale, and not rebuilt.[14]

John Persehouse built a windmill at Walsall in the second decade of the 17th Century. The present Windmill Street marks its probable site. The mill is shown on Joseph Browne's county map of 1682, but was demolished sometime between 1732 and 1735.[15] Browne's map also shows a second windmill at Walsall. It was built about 1672 by John Blackham, a local baker. The mill was purchased by Sir Thomas Wilbrahim in 1675.[16] Apparently upon the expiration of the lease, this probable postmill, was demolished, and replaced by a brick tower mill. With two pairs of sones, the windmill continued its working life until about 1866. Left to decay, the mill was restored in the late 1920s, and converted into an observatory. It still stands on the high ground between Highgate Road and Sandwell Street. A fourth Walsall windmill is shown on William Yates' Map of Staffordshire, published in 1799. It was still there in 1816, and shown on the Ordnance Survey drawing of that year. It was built west of the Birmingham Road, in the crescent that is now Jesson Road. Interestingly the district wherein this and Highgate Mill were situated, was known as Windmill.

There were two windmills at Bloxwich. They were situated between Great and Little Bloxwich. E.J. Homeshaw states that one of them, situated above the Flaxcroft, was there as early as the 16th Century.[17] In 1609 Sir Richard Wilbrahim mortgaged the "Lordship of the Manor of the Foreign of Walsall (Bloxwich)". Included in the estate was a windmill. By 1682 there were two mills here, as is shown in Joseph Browne's map of that year. Millfield Avenue in Little Bloxwich may mark

Highgate mill, Walsall, c.1970.
Reproduced by courtesy of Walsall Public Libraries.

the site of one of the mills.   Yates' Map of Staffordshire for 1775 shows the second as being just south of the Lichfield Road, and east of the High Street, in Chapel Field.   This mill survived until about 1817.

A postmill at Rushall, described as "new erected" in 1693,[18] is depicted on Yates' Staffordshire map of 1775, and on Sherriff's 1788 map.   This windmill was demolished early in the 19th Century[19] sometime before 1834.   It stood on the eastern side of the Lichfield Road, near the northern junction of Cartbridge Lane. Nearby at Goscote Field, stood another windmill, built sometime between 1735 and 1744.[20]   William Yates' map shows the mill a little west of Goscote Lane.   This would place it somewhere in the circle of roads making up the present Severn, Thames and Mersey Roads.   Possibly a smock or post mill, it is shown on James Sherriff's map of 1788, and another of 1799.   It had however disappeared without trace by 1816, as it is not shown on the 2 inch to 1 mile drawing of that year prepared by the Ordnance Survey.   The last two of Walsall's windmills, two postmills, are shown on Joseph Browne's map of Staffordshire for 1682.   They were situated between Pelsall and Shelfield.   Possibly Medieval in origin, both mills appear to have been demolished by 1693, as they are not mentioned in the glebe terrier of that year.[21]

James Gould in his "Men of Aldridge,"[22] alludes to a windmill at Aldridge in the 17th Century.   He tentatively places it at a tumulus known as Gossy Knob, close to the rectory, on a conjectural plan of the town.   No mill is shown on Browne's map, nor on the later 18th Century maps.

There were three windmills to the west of Walsall, at Willenhall.   The first was a postmill with a brick-built roundhouse, situated at the junction of Mill Lane and Lucknow Road.   Joseph Browne depicts it in 1682 at "Nechels".   To the north of the town centre, it is not shown on any of the 18th Century maps of Staffordshire. Though its sails and wooden superstructure were long removed, the roundhouse, with the remains of the main post, was used as a stable until 1949-50, when it was demolished.[23]   A postmill situated to the east of Rose Hill, Willenhall, was one of two mills standing side by side on what is now the Bilston Road Housing Estate.   Owned by the Molineux family, it is mentioned in a declaration executed by George Robinson, a solicitor, in 1850, concerning the Crockett (formerly Molineux) Estate.[24]

The third of Willenhall's windmills was a known Tudor mill, and possibly Medieval in origin.   An inquiry in the Court Baron of the Manor of Stowheath in 1655, revealed that Francis Day had been miller there for the previous 30 years. He paid an annual rent of 26s.8d; half of which was paid to the vicar of Willenhall. The following year Edward Hill became miller, and he in turn was succeeded by John Perry of Bilston.   Stephen Shaw followed him, and in 1670 sold the mill to Thomas Bayley.

Joseph Browne's map of 1682 shows this postmill, just south of the town. Bayley's heirs sold the mill to William and Rebecca Smith in 1698.   Their son William took over the mill.   He was also to become a miller at Wolverhampton and later at Dudley.   In 1756 he sold the windmill to Joseph Clemson.   The mill and its site are described in a court roll of the Chapelry Estate:

"All that moiety or half part undivided, of the Windmill situated in Hill Field and the moiety of that parcel of land whereon the mill stands containing sixteen yards or thereabouts on every side from the mill post extending upon the land formerly in the possession of John Perry, the other moiety of which said Windmill and land now belonging to Joseph Clemson of Willenhall aforesaid, baker, and the said mill is now in his possession."

Sometime after this date the old postmill was either destroyed or demolished, and replaced with a brick-built tower mill. This mill eventually passed into the hands of Mary, the only child of John Clemson. She sold it and the site to Spittle's Colliery who demolished the mill.[25] The Clemson windmill stood just south of Tyler Road, very close to St. Giles Road, Willenhall.

Mill Street in Darlaston takes its name from the windmill situated near the junction with Alfred Street. Built by 1818, it is shown on the 1 inch to 1 mile Ordnance Survey Map of 1834. Darlaston's millers are recorded up to 1878, whereafter the mill appears to have fallen into disuse. It is referred to as such, on the 6 inch to 1 mile map of 1886, though it may well have been demolished by that date. Frederick William Hackwood's comment confirms its disappearance by the following year. Listing field names in the parish, he writes:
"Mill Fields — This field in the centre of the parish (with a footpath across it from Cock Street to Pinfold Street and to Wolverhampton Lane) fixes the position of the windmill."[26]

The mill house still survives, just off Dorsett Road Terrace.

There was a windmill at Wolverhampton by 1587, and possibly earlier. It was built in Quabb or Quebbe Field, "alias Windmill Field."[27] The mill is shown on Joseph Browne's map of 1682, just north of the town. It was situated in Lower Stafford Street, a short distance from the turn into Cannock Road. William Yates depicts the mill as a tower mill in his map of 1775. It appears on Sherriff's map of 1788, and is shown on the 2 inch to 1 mile drawing of 1816. Due to the heavy development of the area, and smallness of the scale of 1 inch to 1 mile map of 1834, it is impossible to distinguish whether the mill was still standing. The later 6 inch to 1 mile map of 1886 shows a circular building on the site, but no reference is made to it being the mill.

A towermill was situated to the east of Goldthorn Road near Grange Road, Wolverhampton. It is first shown on Greenwood's Map of Staffordshire for 1818, but not on the 2 inch drawings prepared by the Ordnance Survey in 1816. With a working life of nearly 90 years, it is marked as "Old Windmill" on the 6 inch to 1 mile map of 1902. Goldthorn mill was demolished soon after.

To the north-east of Wolverhampton is Essington, on the northern edge of the Black Country. Henry Vernon had a postmill built here in 1681, a fact he recorded by having the date deeply cut into the wood above the doorway.[28] Essington Mill is not shown on Joseph Browne's map of 1682. The survey was undoubtedly carried out before the mill's construction. A round house of red brick was added, and Yates shows Essington's mill between Essington village and Moseley, in his county map of 1775. The mill ended its working life in the 1880s. John Fullwood gives a view of the mill at dusk in his book, "Remnants of Old Wolverhampton."[29] The sails, ladder and tailpole had all gone by the outbreak of World War II. With neglect and vandalism Essington windmill has all but disappeared. The remains of the mill are to be seen on a small hillock in Windmill Farm, on the southside of Bognop Road, just

Lower Stafford St. mill, Wolverhampton, c.1870.
Reproduced by courtesy of Wolverhampton Public Libraries.

Goldthorn mill, Wolverhampton c.1902, minus its cap and sails.
Reproduced by courtesy of Wolverhampton Public Libraries.

over the border from Wolverhampton.   Another windmill which has fared a little better than Essington mill, is the nearby Tettenhall mill.   A squat red brick towermill, two thirds of a mile north-east of Wightwick, it was built sometime between 1816 and 1820.   The Ordnance Survey 6 inch to 1 mile map of 1886 has the mill marked as "Windmill House", which suggests that it may have ended its working life.   The mill today stands on high ground just east of Mill Lane, in Mill Close.   Two windmills are shown at Wightwick on William Yates' map of Staffordshire for 1775.   They straddle Windmill Lane.   The mill north of the lane was situated in the present Windmill Crescent.   This windmill has now disappeared completely.   The second mill, which has survived, is situated at the junction of Headland Road and Ashenden Rise.   One of the bricks on the ground floor is inscribed, "This mill was built by John Chamber L in 1720."   The mill ceased its working life sometime during the 1880s, and was converted into a dwelling.   In 1949 Sir Geoffrey Mander presented the mill to Tettenhall Urban District Council.[30]   Another of Wolverhampton's early windmills was situated at Penn.   In a manorial survey carried out at Penn on 18th August 1647,[31] a piece of land, "One butt at Windmill Hill," is recorded at Over Penn.   The hill measured one rood, and was valued at one shilling and sixpence.   The mill had evidently been demolished by 1682, as Penn Windmill is not shown on Joseph Browne's map of Staffordshire for that year.

Several gentlemen at Bilston in 1458 gave some of their lands for the founding of the Chantry of St. Leonard.   One of the fields donated was "Windmill Field."[32] A new windmill, or its successor, is shown as a postmill on Browne's map, and in 1699 one of Bilston's millers is revealed in "The halfe yeare's Chiefe Rent Booke due at Lady Day. . ."[33]   The entry reads:

"Jon Jesson for houses bought of Price (and for land whereon his Windmill stands)."

By 1771 the mill appears to have been demolished.   A survey of Bilston carried out that year records Field No. 814 as "Windmill Croft."[34]   It was occupied by Thomas Tomkys, and measured 1 acre and 8 perchs.   There is no mention of the mill itself though, and no windmill is shown on William Yates' map of Staffordshire for 1775.   A towermill was built to replace the old postmill, at Mount Pleasant.   It was in existence by 1791, and contained two pairs of French stones.   Late in life it worked with only two sails, before being converted into a steam mill to produce cement.   At the time of its conversion from wind to steam some of the wood from the mill was made into a table and two reading desks which found their way into the local Wesley Church.[35]   The mill ceased its working life in the 1870s, and becoming unsafe, was demolished in the early 1970s.

Frederick William Hackwood, Wednesbury's local historian, was of the opinion that the town possessed three windmills by the 14th Century.[36]   The first proof of a mill here appears in 1582.   In an action brought against William Comberford, Thomas Parkes claimed that Comberford, "Insists that all corn be ground at his own mill — and there is only one windmill, built within the last thirty years and insufficient for the needs of a tenth of the inhabitants."[37]

By 1682 there were three postmills at Wednesbury, as is shown on Browne's map for that year.   Robert Plot mentions the two mills in the town:

Essington Windmill

". . . the ingenious Mr.

Millar (Vicar of Wednesbury)" told Plot that there was "a very distinct echo when the windmill window stands open towards the church, otherwise none at all, two of the three windmills there answering the five bells orderly and distinctly."[38]

The two mills referred to stood near the top of the present Squires Walk, and at the top of Windmill Street. The third mill was situated in the northern part of the parish at King's Hill, just east of the present Park Road. All three mills had been demolished by 1886.

West of Wednesbury at Coseley, were two windmills. The first was situated at what is now Denise Drive, just off Ivyhouse Lane, Rosehill. It is shown on the 2 inch to 1 mile Ordnance Survey drawing of 1816, and on the later 1 inch to 1 mile map of 1834, where it is labelled, "Coseley Mill." It was demolished sometime before 1886. The second is a brick-built tower mill. Originally with four sails, boat cap and fan-tail gear mechanism, it was built about 1780. The Jurers' Book of 1784 lists the miller as Joseph Maullin.[39] For several generations the family remained as millers, before being replaced by the King family. They in turn held it until 1871. Soon after the mill fell into disuse. By 1895 both its sails and machinery had been removed.[40] Coseley windmill was converted into a house. It is situated on the hill top to the south of Oak Street. Coseley new mill is depicted in the memorial East window of St. Chad's Church which stands nearby.

Ruiton Windmill, Upper Gornal

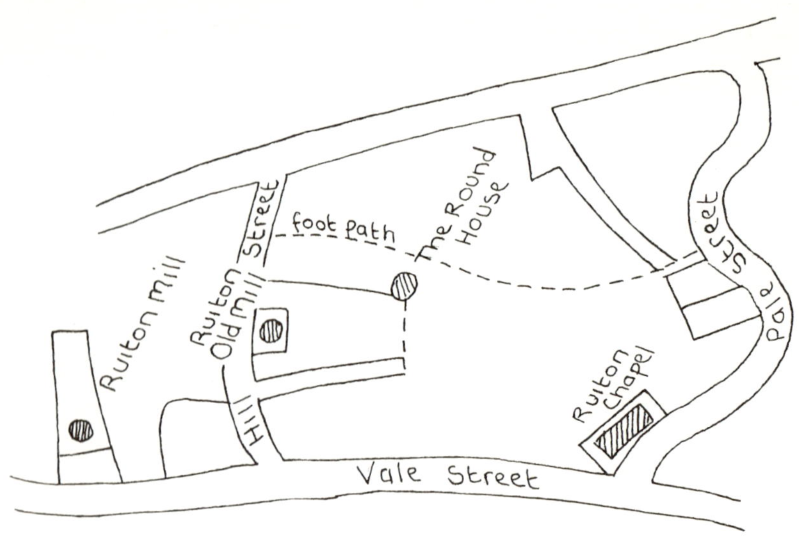

At Sedgley there were four windmills.   Three of them were at Ruiton, and one at Sandyfields.   Ruiton Old Mill is shown on Browne's 1682 map, situated at "Gournall."   It was the property of the Persehouse family.   By an indenture dated 22nd August 1702, Peter Persehouse sold , " all that corn or windmill, built with stone called Rewardine Field, in the parish of Sedgley, adjoining to the road-way there leading from Over Gornall to Nether Gornall, together with the mill, ring, or circle of ground on which the said mill standeth. . . to Thomas Maullin, the elder of Coseley, Nailor. . ."

The mill remained in the possession of the Maullin family until 1759, when Job Maullin sold it to two local speculators.   They sold it the next year to Samuel Fereday, who not being a miller himself, leased it to William Turton and later Thomas Turner.   It was while Turner was miller here that Fereday's son, Samuel jr., sold it to him and his partner Benjamin Lloyd.   Lloyd for a time worked alone as miller, before selling Ruiton Old Mill to William Richmond, millwright at Himley.   In 1835 he sold it to the Gornall miller, George Richmond.   The business failed, and three years later the bankrupt Richmond had the mill seized in lieu of undishcarged debts.   It passed into the hands of the Wolverhampton bankers, Holyoake.   For a time they leased the mill before selling it to Paul Thompson.   In 1861 he sold it to Edward Jones.   On Wednesday, 31st. January 1872 the Old Mill, situated at the corner of Hill Street and Windmill Street, fell to the ground with a tremendous crash.   Its fall was attributed to vibrations from the nearby stone quarry and prolonged rainfall.

Ruiton tower mill.
Reproduced by courtesy of Frank Power of Dudley, photographer.

Smethwick Windmill, as part of Cheshire's Brewery

Ruiton New Mill was situated in Vale Street Ruiton. This yellow sandstone tower mill was built for George Richmond about 1830. He was also owner of the Old Mill referred to above. After 1838 the mill passed through the hands of several owners before ceasing its working life about 1880. Its stocks and much of its machinery survived until 1958. Ruiton New Mill is now in the care of the Black Country Society.

The third windmill at Ruiton was situated at the lower end of Windmill Street. It was built about 1830, and appears on the 1 inch to 1 mile Ordnance Survey Map of 1834. Built of local sandstone, this tower mill was used for grinding stone into sand for the building trade. About the turn of the century the mill was converted into a house, and was known locally as the "Round House". It became uninhabitable and fell into decay. The mill collapsed in March 1961.

Sedgley's fourth windmill was situated at the junction of Sandyfields Road and Oakleigh Drive. A postmill with round house, it was drawn by the local artist R. Noyes in 1817. Sandyfields mill is shown on both the Ordnance Survey's 2 inch to 1 mile drawings of 1816, and the 1834 Map. Its last miller was Abel Fletcher, who worked the mill from 1850 to 1872 or a little later. There are no further directory entries after 1872, and the mill is not shown on the 6 inch to 1 mile Ordnance Survey map of 1886.

A postmill is shown at Tipton on Joseph Browne's Map of 1682. In an indenture, "bearing date 14th and 15th days of September 1768," both the windmill and a watermill are referred to.[41] However by 1792 the windmill appears to have been demolished. In that year a number of trustees purchased land adjoining Upper and Lower Church Lane, with the intention of building a new church dedicated to St. Martin. Among the land they acquired was "Windmill Piece."[42] There is no mention of the actual mill. As Yates' map of 1775 does not show Tipton Windmill it would appear that the mill was demolished sometime between 1768 and 1775.

Dudley had three windmills. The oldest, possibly the Manor windmill, was in existence by the early 16th Century,[43] and may have been much older. The mill, or more probably its successor, is shown as a tower, or smock mill on a map of Dudley, c. 1700,[44] and on Isaac Taylor's County Map of Worcestershire for 1802. It is shown to be situated at a place called "Paradise," which coincides with the present triangle of roads that make up Stafford, Albert and Steppingstone Streets. Due to the heavy development of the area soon after, the mill is not apparently shown on the Ordnance Survey drawing of the area for 1816, nor the subsequent map of 1834.

A short-lived Dudley windmill was situated between the present Salop and King Edmund Streets, almost facing West Street. It is shown as an apparent tower or smock mill on a map of Dudley, c. 1700, apparently having been built post - 1682. It was either destroyed or demolished by 1787.[45].

The last, was Shavers End Mill, situated to the north of Windmill Street, facing St. James' Terrace. The mill is shown on Browne's county map of 1682. Throughout the 18th Century the Bagley family owned the mill. Dud Bagley is recorded as miller in 1787.[46] Shavers Mill appears on all of the 18th and early 19th Century maps, including the Ordnance Survey drawing of 1816. Between then and the subsequent 1834 map, the windmill was demolished. The site is there marked as "Iron-Stone Mines."

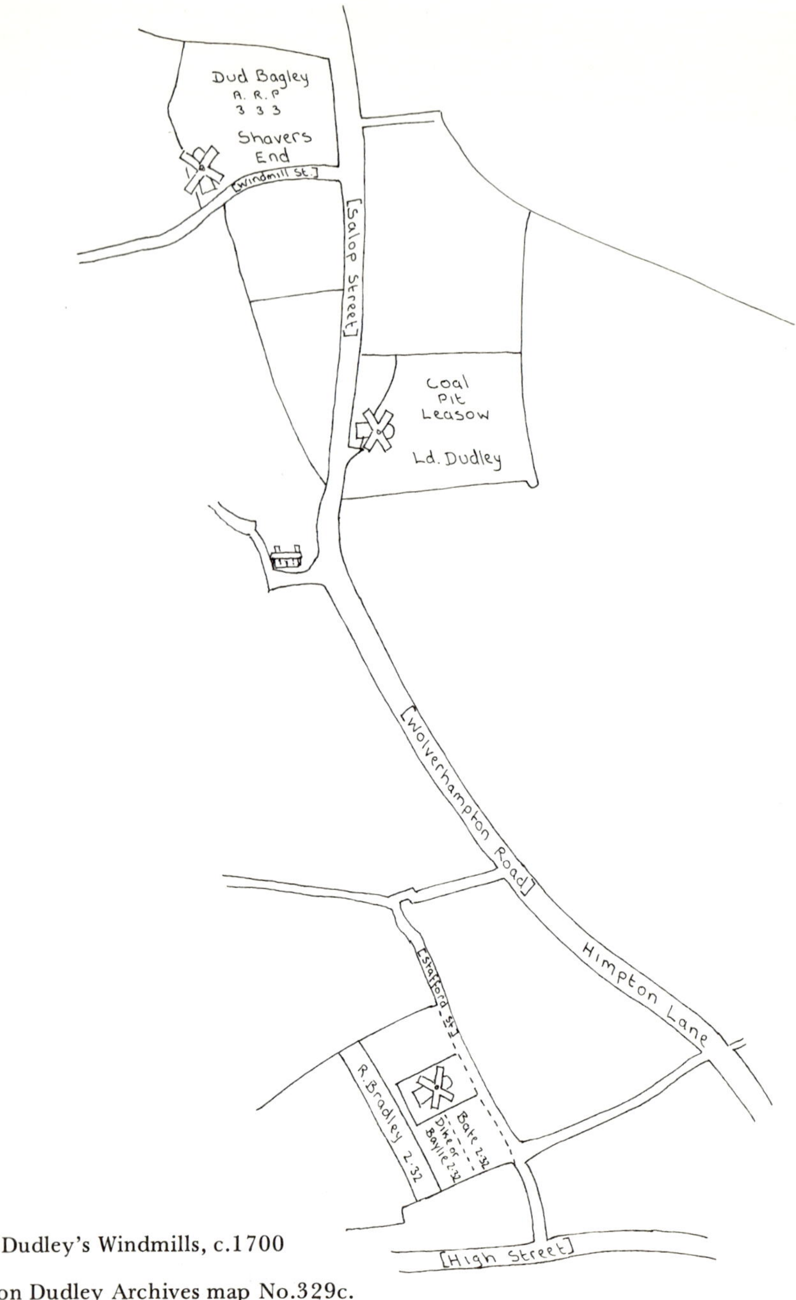

Dudley's Windmills, c.1700

based on Dudley Archives map No.329c.
Modern roads appear in brackets.

26

West of Dudley, in the parish of Kingswinford, were two windmills, Kingswinford Mill and Tansey Green Mill.   Kingswinford Mill was built about 1818.   A red-brick tower mill it is recorded in William Fowler's survey of the parish in 1824,[47] and shown on the 1 inch to 1 mile Ordnance Survey map of 1834.   Situated in the western part of the parish, just south of Wall Heath, the mill was converted into a summerhouse at the turn of the century.[48]

Tansey Green mill was situated in the eastern part of the parish.   Fowler in his survey records Field No. 2168 as "House, Windmill & Yard."   The miller's name is given as William Downing.   In a later survey of 1840, William Meredith is recorded as miller.[49]   The Tile Works at the junction of Dreadnaught and Tansey Roads, Kingswinford, mark the site of this windmill.

North-east of Stourbridge stood Wollaston Windmill.   It was situated near the junction of Vicarage and Gerald Roads, facing Firmstone Street.   It is not recorded on Browne's 1682 map, but was built by 1699.   On Josiah Bach's map of the Parish of Old Swinford, drawn that year, the mill is shown, and marked as "Old Windmill."   This would tend to suggest that it was in existence by 1682, but just omitted from Browne's map.   The mill stood in a field known as "Windmill Piece."   Bach gives the owner as Mr. Foley.   It, or possibly a successor, is shown on the 2 inch to 1 mile Ordnance Survey drawing of 1816, and on the 1834 map.   Wollaston Mill is not recorded on 6 inch O.S. map of 1884, and was presumably demolished by that date. Stourbridge itself had a windmill, situated within the triangle of roads made up by Church Street and Hagley and Cranbie Roads.   James Sherriff shows the mill just south of the town centre, on his map of 1788.[50]   The mill is shown on the 2 inch O.S. drawing of 1814, but not on the completed map of 1834.   Strangely, Isaac Taylor, while showing nearby Wollaston Mill on his map of 1802, does not show Stourbridge Mill.

Himley, situated on the edge of the Black Country, had a windmill by 1682. Browne shows it just north of the village on his county map.   It is not shown on either Yates' map of 1775, nor Sherriff's map of 1788, which tends to suggest that the mill had ceased to exist.   What is probably a replacement windmill is shown west of the village, south of Himley Common, on the 1 inch to 1 mile Ordnance Survey map of 1834.   The mill was either destroyed or demolished by 1886.

Moving south-east now to Cradley, where James Sherriff depicts two post-mills on his map of 1788.   They were situated on Windmill Hill, between the present Two Gates and Toy's Lanes.   The mills are shown on the 2 inch drawing of 1816, and on the 1834 Ordnance Survey map.   Both windmills had been demolished by 1885.

South to Halesowen and Quarry Hill.   There were two windmills here, just south of the Hagley Road.   Both are shown on Sherriff's map of 1788, but by 1816 one of the mills had been demolished.   The remaining mill is recorded on the 1 inch map of 1834, but it too had gone by 1885.

Finally the last of the known Black Country windmills, Rowley Regis mill. Situated just north of the town, Browne shows its 1682 site as lying between "Iberty" and "Portway".   The mill is shown on Yates' map of 1775.   He also records a Wind-mill Lane at the western edge of the parish near its boundary with Netherton.   A map of Rowley, drawn about 1826[51] has the windmill clearly marked at Tipity Green.

Its present site would be just off Hawes Lane, near Club Buildings.   Again at the western edge of the parish there is a further indication of a windmill at Windmill End. No mill is shown here though.   Rowley Mill was demolished before 1887.

1.   Smethwick Telephone, Sat. 13th Aug. 1949
2.   Birmingham Weekly Post, Sat. 16th. Jan. 1932
3.   Staffs Historical Collection, New Series VI (1) p.13
4.   B.R.L. 297302
5.   West Bromwich, 1882 p.123
6.   B.R.L. 457177
7.   It is not shown on Joseph Browne's county map of that year.
8.   History of West Bromwich by Mary Willett, 1882 p.178
9.   R. Whitworth, Plan of Intended Navigation from Birmingham. . . 1767
10.   Yates' Map of Staffs. 1775.
11.   Chartulary 33 Edward I, quoted in "History of the Borough and Foreign of Walsall," by E.L. Glew, 1856 p.26
12.   A History of Walsall by Frederick Willmore, 1887 p.65-6
13.   Walsall Records p.39
14.   V.C.H. Staffs vol VIII, p.186
15.   Lichfield Joint Record Office B/V/61 Walsall, 1732 and 1735
16.   Staffs Record Office D.1287/6/14
17.   The Story of Bloxwich, 1955, p.57
18.   Terrier of Church Lane, 1693
19.   Records of Rushall by Frederick W. Willmore, 1892, p.110
20.   Lichfield Joint Record Office, Walsall Terriers B/V/6/Walsall
21.   Ibid (1693)
22.   p.15 & 60
23.   A History of Willenhall by Norman W. Tildesley, 1951 p.180
24.   Ibid p.181
25.   A History of Willenhall by Norman W. Tildesley, 1951 p.181-2
26.   History of Darlaston, 1887 p.13
27.   Gerald P. Mander, "A History of Wolverhampton, 1960 p.9 & 44
28.   Staffordshire LIfe & County Pictorial vol 6. No.1 Spring 1954 p.24
29.   Published 1880, plate 9
30.   Windmills in Staffordshire by W.A. Seaby & A.C. Smith, 1979 p.23
31.   "Investigating Penn" 1975 p.48 Ed. T.R. Bennett
32.   Rotuli patents for the 35th year Henry VI. Quoted in "History of Bilston" by G.T. Lawley, 1893 p.24-5
33.   "Bilston in the 17th Century" by G.T. Lawley, 1920 p.62
34.   "An Historical Account of Bilston" by Joseph Price, 1835 p.86
35.   Birmingham Weekly Post 25th June 1932
36.   Wednesbury Ancient & Modern, 1902, p.54
37.   P.R.O. c3/221/147 (1582)
38.   "Natural History of Staffordshire" 1686
39.   "History of Coseley" by J.S. Roper, 1952, p.65
40.   Seaby & Smith "Windmills in Staffordshire", 1979, p.17
41.   "History of Tipton" by John Parkes, 1915, p.76-7
42.   Ibid
43.   Dudley Archives, Rent Rolls 19/36
44.   Ibid. Map No.329c.
45.   It is not shown on Dudley Parish Map. 1787. Dudley Archives Map No.892A

46. Ibid
47. B.R.L. 69168
48. Seaby & Smith p.19
49. B.R.L. 598722
50. A map of upwards of 25 miles around the Town of Birmingham, 1788. (Survey completed in 1796)
51. B.R.L. 396416A

A GAZETTEER OF WINDMILLS IN BIRMINGHAM
AND THE BLACK COUNTRY

Aldridge:	Aldridge Mill, a postmill.
Bilston:	Bilston Mill, a postmill Mount Pleasant Mill, a brick-built tower mill.
Birmingham:	Baskerville's mill, small brick-built tower mill. Chapman's mill, postmill and later a brick-built tower mill. Cooper's mill, a smock mill. Heath mill, a smock mill.
Bloxwich:	Little Bloxwich mill, a postmill. Chapel Field mill, (?) postmill.
Coseley:	Coseley Old mill, a postmill. Coseley New mill, a tower mill.
Cradley:	Cradley mills, postmills.
Darlaston:	Darlaston mill, a postmill.
Dudley:	Paradise (or the Manor) mill, postmill and later a smockmill. Salop St. mill (?) smockmill. Shavers End mill, a post and later a smockmill.
Edgbaston:	Speedwell mill, a postmill.
Erdington:	Slade mill, a postmill.
Essington:	Essington mill, a postmill.
Handsworth:	Bristnalls End Mill, (?) postmill. Hamstead mill, a postmill. Hutton's mill, a kiln-shaped brick tower mill. Soho mill (?) a postmill.
Himley:	Old Himley mill, a postmill. Himley mill, (?) a smockmill.
Kingswinford:	Kingswinford mill, a redbrick tower mill. Tansey Green mill (?) a postmill.
Northfield:	Middleton Hall mill, a postmill. Shenley mill, a postmill.
Pelsall:	Pelsall mill, a postmill.
Penn:	Penn mill, a postmill.
Perry Barr:	Holdford mill, a post or smock mill.
Quarry Hill:	Quarry Hill mills (?) smockmills.
Quinton:	Ridgacre mill (?) a postmill.
Rowley Regis:	Rowley mill, a postmill.
Rushall:	Goscote mill, (?) a smockmill. Rushall mill, a postmill.

Saltley:	Saltley mill, a postmill.
Sedgley:	"Roundhouse" mill, a sandstone tower mill. Ruiton Old mill (or Gornal mill), a postmill. Ruiton New mill, a sandstone tower mill. Sandyfields mill, a postmill.
Sheldon:	Lyndon Green mill, a postmill. Sheldon mill, a postmill. Wells Green, a postmill.
Shelfield:	Shelfield mill, a postmill.
Smethwick:	Smethwick mill, a brick-built tower mill.
Stourbridge:	Stourbridge: Stourbridge mill, (?) a smockmill.
Sutton Coldfield:	High Heath mill, a postmill. Langley mill, (?) a postmill. Maney Hill, a postmill.
Tettenhall:	Tettenhall mill, a squat redbrick towermill.
Tipton:	Tipton mill, a postmill.
Walsall:	Caldmore mill, a postmill. Highgate mill, a postmill later replaced by a tower mill. Persehouse mill, a postmill. Walsall mill, a postmill.
Wednesbury:	Comberford's mill, a postmill. Kings Hill mill, a postmill. Wednesbury mill, a postmill.
West Bromwich:	Hall Green mill, a postmill. Lyndon Mill, a postmill. Ryder's mill, a postmill. Tantany mill, a postmill.
Wightwick:	Wightwick mills, two brick-built tower mills.
Willenhall:	Nechels mill, a postmill. Rose Hill mill, a post mill. Stowheath mill, a postmill.
Witton:	Bleak Hills mill, (?) a smockmill.
Wollaston:	Wollaston mill, (?) a smockmill.
Wolverhampton:	Goldthorn Hill mill, a brick-built tower mill. Quabb Field (or Lower Stafford St.) mill, a postmill, replaced by a brick-built tower mill.
Yardley:	Red Hill mill, a postmill. Wake Green mill, a postmill. Yardley Wood mill, a postmill.